# 100 facts on
# Space

# 100 facts on Space

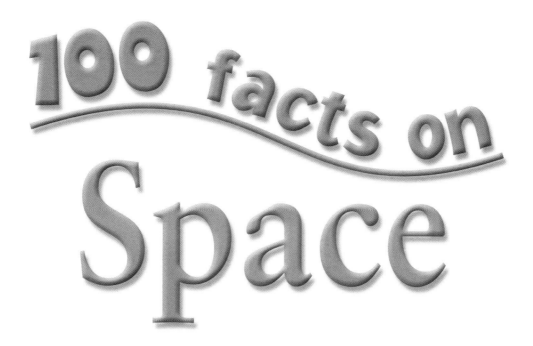

Sue Becklake

Consultant: Peter Bond

Miles Kelly

First published as hardback in 2002 by Miles Kelly Publishing Ltd
Harding's Barn, Bardfield End Green, Thaxted, Essex, CM6 3PX, UK

Copyright © Miles Kelly Publishing Ltd 2002

This edition printed 2010

2 4 6 8 10 9 7 5 3 1

**Editorial Director:** Belinda Gallagher
**Art Director:** Jo Brewer
**Assistant Editor:** Lucy Dowling
**Volume Designer:** John Christopher, White Design
**Proofreader/Indexer:** Lynn Bresler
**Production Manager:** Elizabeth Collins
**Reprographics:** Anthony Cambray, Liberty Newton, Ian Paulyn
**Assets Manager:** Bethan Ellish

ISBN 978-1-84236-760-5

Printed in China

British Library Cataloguing-in-Publication Data
A catalogue record for this book is available from the British Library

**ACKNOWLEDGEMENTS**
The publishers would like to thank the following artists
who have contributed to this book:
Kuo Kang Chen, Alan Hancocks, Janos Marffy,
Martin Sanders, Mike Saunders, Rudi Vizi

Cartoons by Mark Davis at Mackerel

All other images from Miles Kelly Archives

Made with paper from a sustainable forest

www.mileskelly.net
info@mileskelly.net

www.factsforprojects.com

# Contents

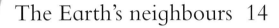

# Surrounded by space

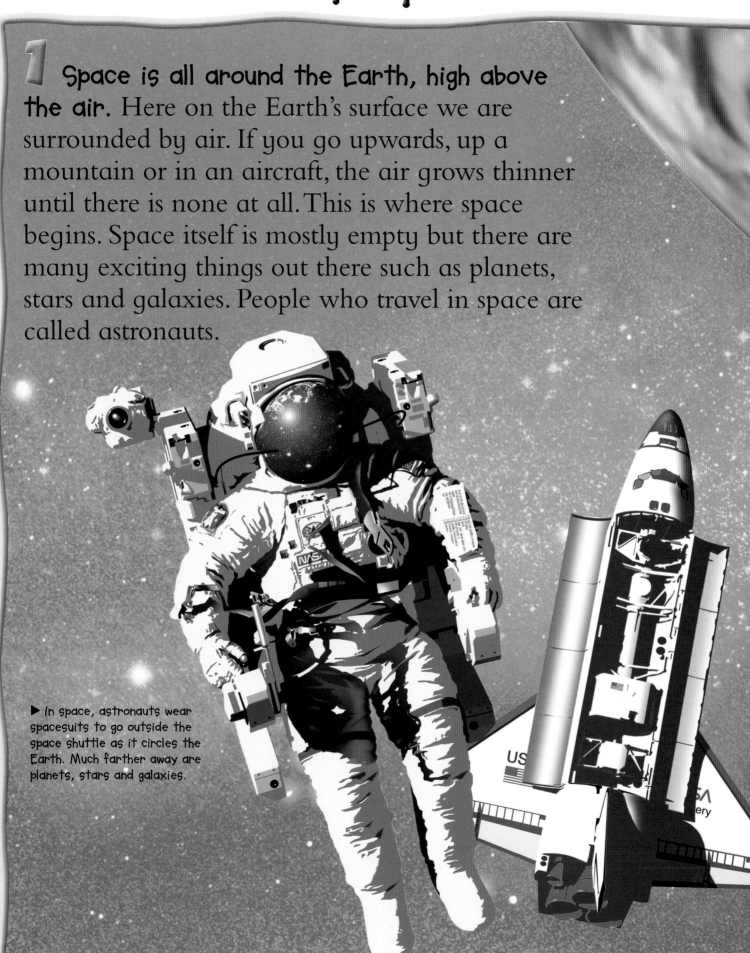

**1** **Space is all around the Earth, high above the air.** Here on the Earth's surface we are surrounded by air. If you go upwards, up a mountain or in an aircraft, the air grows thinner until there is none at all. This is where space begins. Space itself is mostly empty but there are many exciting things out there such as planets, stars and galaxies. People who travel in space are called astronauts.

▶ In space, astronauts wear spacesuits to go outside the space shuttle as it circles the Earth. Much farther away are planets, stars and galaxies.

# Our life-giving star

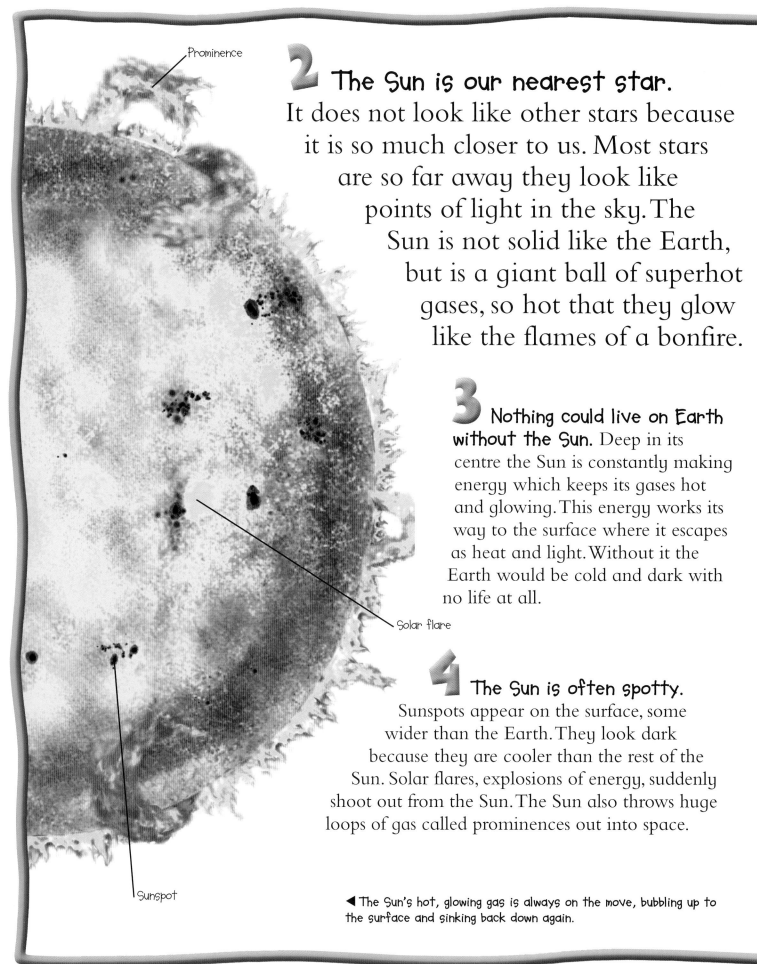

Prominence

**2** **The Sun is our nearest star.** It does not look like other stars because it is so much closer to us. Most stars are so far away they look like points of light in the sky. The Sun is not solid like the Earth, but is a giant ball of superhot gases, so hot that they glow like the flames of a bonfire.

**3** **Nothing could live on Earth without the Sun.** Deep in its centre the Sun is constantly making energy which keeps its gases hot and glowing. This energy works its way to the surface where it escapes as heat and light. Without it the Earth would be cold and dark with no life at all.

Solar flare

**4** **The Sun is often spotty.** Sunspots appear on the surface, some wider than the Earth. They look dark because they are cooler than the rest of the Sun. Solar flares, explosions of energy, suddenly shoot out from the Sun. The Sun also throws huge loops of gas called prominences out into space.

Sunspot

◄ The Sun's hot, glowing gas is always on the move, bubbling up to the surface and sinking back down again.

# 5 When the Moon hides the Sun there is an eclipse.

Every so often, the Sun, Moon and Earth line up in space so that the Moon comes directly between the Earth and the Sun. This stops the sunlight from reaching a small area on Earth. This area grows dark and cold, as if night has come early.

▼ When the Moon casts a shadow on the Earth, there is an eclipse of the Sun.

▶ When there is an eclipse, we can see the corona (glowing gas) around the Sun.

Sun

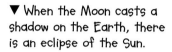

**WARNING:**
Never look directly at the Sun especially through a telescope or binoculars. It is so bright it will harm your eyes or even make you blind.

## I DON'T BELIEVE IT!

The surface of the Sun is nearly 60 times hotter than boiling water. It is so hot it would melt a spacecraft flying near it.

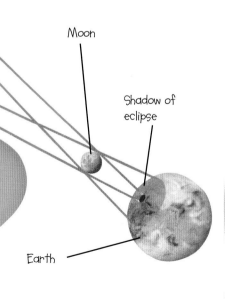

Moon

Shadow of eclipse

Earth

# A family of planets

**6** **The Sun is surrounded by a family of circling planets called the Solar System.** This family is held together by an invisible force called gravity, which pulls things towards each other. It is the same force that pulls us down to the ground and stops us from floating away. The Sun's gravity pulls on the planets and keeps them circling around it.

**7** **The Earth is one of eight planets in the Sun's family.** They all circle the Sun at different distances from it. The four planets nearest to the Sun are all balls of rock. The next four planets are much bigger and are made of gas and liquid. The tiny dwarf planet at the edge of the Solar System, Pluto, is a solid, icy ball.

**8** **Moons circle the planets, travelling with them round the Sun.** Earth has one Moon. It circles the Earth while the Earth circles round the Sun. Pluto also has one moon. Mars has two tiny moons but Mercury and Venus have none at all. There are large families of moons, like miniature solar systems, around all the large gas planets.

Saturn

Uranus

Neptune

Pluto
(dwarf planet)

Sun

Mercury

Moon

Venus

Earth

Mars

Jupiter

▲ The planets are all different. Mercury, nearest the Sun, is small and hot. Then Venus, Earth and Mars are rocky and cooler. Beyond them Jupiter, Saturn, Uranus and Neptune are large and cold, while dwarf planet Pluto is tiny and icy.

**9** There are millions of smaller members in the Sun's family. Some are tiny specks of dust speeding through space between the planets. Larger chunks of rock, many as large as mountains, are called asteroids. Comets come from the edge of the Solar System, skimming past the Sun before they disappear again.

**I DON'T BELIEVE IT !**

If the Sun was the size of a large beach ball, the Earth would be as small as a pea, and the Moon would look like a pinhead.

11

# Planet of life

**10** **The planet we live on is the Earth.** It is a round ball of rock. On the outside where we live the rock is hard and solid. But deep below our feet, inside the Earth, the rock is hot enough to melt. You can sometimes see this hot rock showering out of an erupting volcano.

Outer core

Inner core

**11** **The Earth is the only planet with living creatures.** From space the Earth is a blue and white planet, with huge oceans and wet masses of cloud. People, animals and plants can live on Earth because of all this water.

**12** **Sunshine gives us daylight when it is night on the other side of the Earth.** When it is daytime, your part of the Earth faces towards the Sun and it is light. At night, your part faces away from the Sun and it is dark. Day follows night because the Earth is always turning.

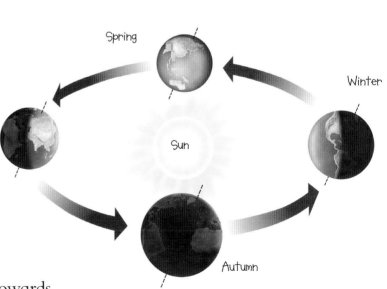

Spring

Winter

Summer

Sun

Autumn

▲ The Earth tilts, so we have different seasons as the Earth moves around the Sun. These are the seasons for the northern half of the Earth.

◄ The inner core at the centre of the Earth is made of iron. It is very hot and keeps the outer core as liquid. Outside this is the mantle, made of thick rock. The thin surface layer that we live on is called the crust.

Crust

Mantle

## I DON'T BELIEVE IT!

The Moon has no air or water. When astronauts went to the Moon they had to take air with them in their spacecraft and space suits.

New Moon

Crescent Moon

First quarter Moon

Gibbous Moon

Full Moon

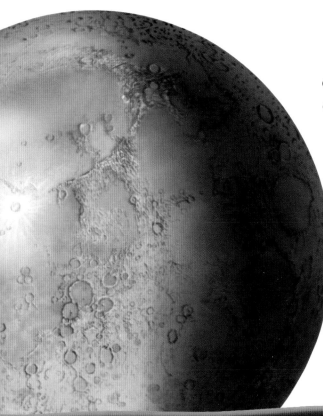

**13** Look for the Moon on clear nights and watch how it seems to change shape. Over a month it changes from a thin crescent to a round shape. This is because sunlight is reflected by the Moon. We see the full Moon when the sunlit side faces the Earth and a thin, crescent shape when the sunlit side is facing away from us.

**14** Craters on the Moon are scars from space rocks crashing into the surface. When a rock smashes into the Moon at high speed, it leaves a saucer-shaped dent, pushing some of the rock outwards into a ring of mountains.

# The Earth's neighbours

**15** Venus and Mars are the nearest planets to the Earth. Venus is closer to the Sun than the Earth while Mars is farther away. Each takes a different amount of time to circle the Sun and we call this its year. A year on Venus is 225 days, on Earth 365 days and on Mars 687 days.

▲ All we can see of Venus from space are the tops of its clouds. They take just four days to race right around the planet.

**16** Venus is the hottest planet. It is hotter than Mercury, although Mercury is closer to the Sun and gets more of the Sun's heat. Heat builds up on Venus because it is completely covered by clouds which trap the heat, like the glass in a greenhouse.

**17** Venus has poisonous clouds with drops of acid that would burn your skin. They are not like clouds on Earth, which are made of droplets of water. These thick clouds do not let much sunshine reach the surface of Venus.

▼ Under its clouds, Venus has hundreds of volcanoes, large and small, all over its surface. We do not know if any of them are still erupting.

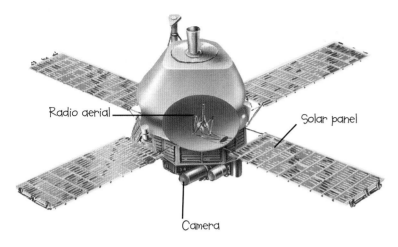

Radio aerial

Solar panel

Camera

## 19

**Winds on Mars whip up huge dust storms that can cover the whole planet.** Mars is very dry, like a desert, and covered in red dust. When a space probe called *Mariner 9* arrived there in 1971, the whole planet was hidden by dust clouds.

◄ *Mariner 9* was the first space probe to circle another planet. It sent back over 7000 pictures of Mars showing giant volcanoes, valleys, ice caps and dried-up river beds.

## 18

**Mars has the largest volcano in the Solar System.** It is called Olympus Mons and is three times as high as Mount Everest, the tallest mountain on Earth. Olympus Mons is an old volcano and it has not erupted for millions of years.

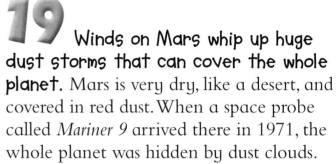

### PLANET-SPOTTING

See if you can spot Venus in the night sky. It is often the first bright 'star' to appear in the evening, just above where the Sun has set. Because of this we sometimes call it the 'evening star'.

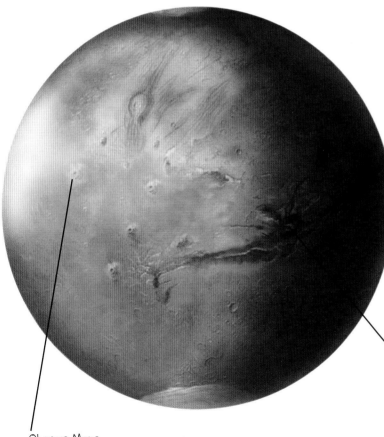

Olympus Mons

## 20

**There are plans to send astronauts to Mars but the journey would take six months or more.** The astronauts would have to take with them everything they need for the journey there and back and for their stay on Mars.

Valles Marineris

◄ An enormous valley seems to cut Mars in half. It is called Valles Marineris. To the left is a row of three huge volcanoes and beyond them you can see the largest volcano, Olympus Mons.

# The smallest of all

**21** Tiny Pluto is so far away, it was not discovered until 1930. In 2006, Pluto was classed as a dwarf planet. It is less than half the width of the smallest planet, Mercury. In fact Pluto is smaller than our Moon.

▲ Pluto is too far away to see any detail on its surface, but it might look like this.

**22** Dwarf planet Pluto is further from the Sun than the eight main planets. If you were to stand on Pluto's surface, the Sun would not look much brighter than the other stars. Pluto gets little heat from the Sun and is completely covered with ice.

**23** Space probes have not yet visited Pluto. So astronomers will have to wait for close-up pictures and detailed information that a probe could send back. Even if one was sent to Pluto it would take at least eight years to get there.

**24** No one knew Pluto had a moon until 1978. An astronomer noticed what looked like a bulge on the side of the planet. It turned out to be a moon and was named Charon. Charon is about half the width of Pluto.

▼ If you were on Pluto, its moon Charon would look much larger than our Moon does, because Charon is very close to Pluto.

# 25

Mercury looks like our Moon. It is a round, cratered ball of rock. Although a little larger than the Moon, like the Moon it has no air.

## MAKE CRATERS
### You will need:
flour    baking tray
a marble or a stone

Spread some flour about 2 centimetres deep in a baking tray and smooth over the surface. Drop a marble or a small round stone onto the flour and see the saucer-shaped crater that it makes.

◀ Mercury's many craters show how often it was hit by space rocks. One was so large that it shattered rocks on the other side of the planet.

▼ The Sun looks huge as it rises on Mercury. A traveller to Mercury would have to keep out of its heat.

# 26

The sunny side of Mercury is boiling hot but the night side is freezing cold. Being the nearest planet to the Sun the sunny side can get twice as hot as an oven. But Mercury spins round slowly so the night side has time to cool down, and there is no air to trap the heat. The night side becomes more than twice as cold as the coldest place on Earth – Antarctica.

# The biggest of all

**27** Jupiter is the biggest planet, more massive than all the other planets in the Solar System put together. It is 11 times as wide as the Earth although it is still much smaller than the Sun. Saturn, the next largest planet, is more than nine times as wide as the Earth.

**28** Jupiter and Saturn are gas giants. They have no solid surface for a spacecraft to land on. All that you can see are the tops of their clouds. Beneath the clouds, the planets are made mostly of gas (like air) and liquid (water is a liquid).

▼ Jupiter's fast winds blow the clouds into coloured bands around the planet.

**29** The Great Red Spot on Jupiter is a 300 year old storm. It was first noticed about 300 years ago and is at least twice as wide as the Earth. It rises above the rest of the clouds and swirls around like storm clouds on Earth.

▼ There are many storms on Jupiter but none as large or long lasting as the Great Red Spot.

▼ Jupiter's Moon Io is always changing because its many volcanoes throw out new material from deep inside it.

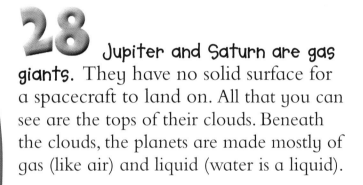

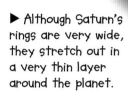

▶ Although Saturn's rings are very wide, they stretch out in a very thin layer around the planet.

**30** **The shining rings around Saturn are made of millions of chunks of ice.** These circle around the planet like tiny moons and shine by reflecting sunlight from their surfaces. Some are as small as ice cubes while others can be as large as a car.

**31** **Jupiter and Saturn spin round so fast that they bulge out in the middle.** This can happen because they are not made of solid rock. As they spin their clouds are stretched out into light and dark bands around them.

**I DON'T BELIEVE IT!**
Saturn is the lightest planet in the Solar System. If there was a large enough sea, it would float like a cork.

**32** **Jupiter's moon Io looks a bit like a pizza.** It has many active volcanoes that throw out huge plumes of material, making red blotches and dark marks on its orange-yellow surface.

# So far away

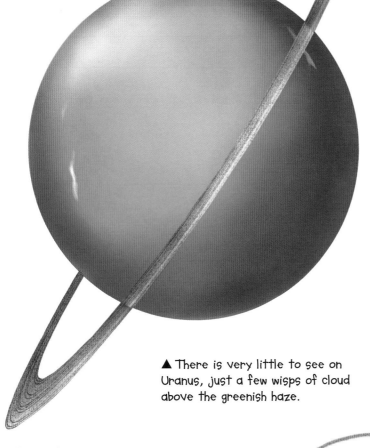

▲ There is very little to see on Uranus, just a few wisps of cloud above the greenish haze.

**33** Uranus and Neptune are gas giants like Jupiter and Saturn. They are the next two planets beyond Saturn but much smaller, being less than half as wide. They too have no hard surface. Their cloud tops make Uranus and Neptune both look blue. They are very cold, being so far from the Sun.

**34** Uranus seems to 'roll' around the Sun. Unlike most of the other planets, which spin upright like a top, Uranus spins on its side. It may have been knocked over when something crashed into it millions of years ago.

**35** Uranus has more moons than any other planet. Twenty-one have been discovered so far, although one is so newly discovered it has not got a name yet. Most of them are very small but there are five larger ones.

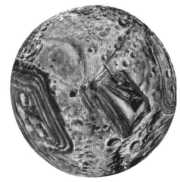

◄ Miranda is one of Uranus' moons. It looks as though it has been split apart and put back together again.

# 36 Neptune had a storm that disappeared.

When the *Voyager 2* space probe flew past Neptune in 1989 it spotted a huge storm like a dark version of the Great Red Spot on Jupiter. When the Hubble Space Telescope looked at Neptune in 1994, the storm had gone.

# 37 Neptune has bright blue clouds that make the whole planet look blue.

Above these clouds are smaller white streaks. These are icy clouds that race around the planet. One of the white clouds seen by the *Voyager 2* space probe was called 'Scooter' because it scooted around the planet so fast.

◄ Like all the gas giant planets, Neptune has rings although they are much darker and thinner than Saturn's rings.

# 38 Neptune is sometimes farther from the Sun than Pluto.

All the main planets travel around the Sun along orbits (paths) that look like circles, but Pluto's path is more squashed. This sometimes brings it closer to the Sun than Neptune.

▼ In the past, astronomers thought there might be another planet, called Planet X, outside Neptune and Pluto.

Orbit of Neptune

Orbit of Planet X

Orbit of Pluto

## QUIZ
1. How many moons does Uranus have?
2. Which is the biggest planet in our Solar System?
3. Which planet seems to 'roll' around the Sun?
4. What colour are Neptune's clouds?

Answers:
1. 21  2. Jupiter
3. Uranus  4. Blue

# Comets, asteroids and meteors

**39** **There are probably billions of tiny comets at the edge of the Solar System.** They circle the Sun far beyond the farthest planet, Neptune. Sometimes one is disturbed and moves inwards towards the Sun, looping around it before going back to where it came from. Some comets come back to the Sun regularly, such as Halley's comet that returns every 76 years.

▲ The solid part of a comet is hidden inside a huge, glowing cloud that stretches into a long tail.

**40** **A comet is often called a dirty snowball because it is made of dust and ice mixed together.** Heat from the Sun melts some of the ice. This makes dust and gas stream away from the comet, forming a huge tail that glows in the sunlight.

**41** **Comet tails always point away from the Sun.** Although it looks bright, a comet's tail is extremely thin so it is blown outwards, away from the Sun. When the comet moves away from the Sun, its tail goes in front of it.

**42** Asteroids are chunks of rock that failed to stick together to make a planet. Most of them circle the Sun between Mars and Jupiter where there would be room for another planet. There are millions of asteroids, some the size of a car, and others as big as mountains.

▶ Asteroids travel in a ring around the Sun. This ring is called the Asteroid belt and can be found between Mars and Jupiter.

**43** Meteors are sometimes called shooting stars. They are not really stars, just streaks of light that flash across the night sky. Meteors are made when pebbles racing through space at high speed hit the top of the air above the Earth. The pebble gets so hot it burns up. We see it as a glowing streak for a few seconds.

## QUIZ
1. Which way does a comet tail always point?
2. What is another name for a meteor?
3. Where is the asteroid belt?

Answers:
1. Away from the Sun
2. Shooting star
3. Between Mars and Jupiter

▼ At certain times of year there are meteor showers when you can see more shooting stars than usual.

# A star is born

**44** **Stars are born in clouds of dust and gas in space called nebulae.** Astronomers can see these clouds as shining patches in the night sky, or dark patches against the distant stars. These clouds shrink as gravity pulls the dust and gas together. At the centre, the gas gets hotter and hotter until a new star is born.

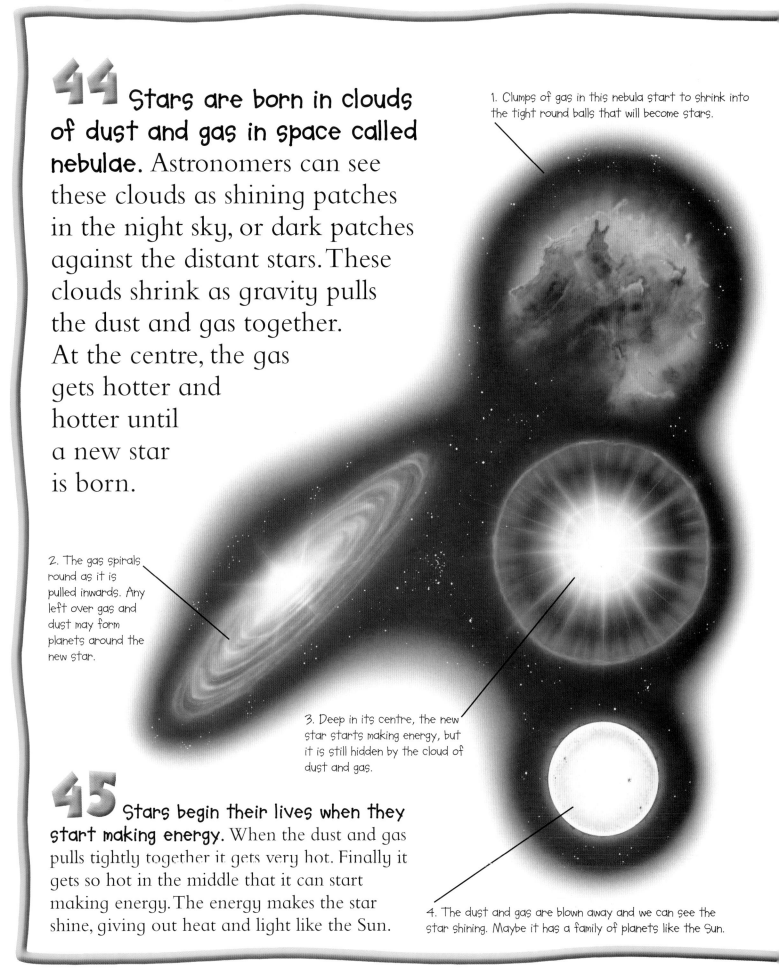

1. Clumps of gas in this nebula start to shrink into the tight round balls that will become stars.

2. The gas spirals round as it is pulled inwards. Any left over gas and dust may form planets around the new star.

3. Deep in its centre, the new star starts making energy, but it is still hidden by the cloud of dust and gas.

4. The dust and gas are blown away and we can see the star shining. Maybe it has a family of planets like the Sun.

**45** **Stars begin their lives when they start making energy.** When the dust and gas pulls tightly together it gets very hot. Finally it gets so hot in the middle that it can start making energy. The energy makes the star shine, giving out heat and light like the Sun.

# 46

**Young stars often stay together in clusters.** When they start to shine they light up the nebula, making it glow with bright colours. Then the starlight blows away the remains of the cloud and we can see a group of new stars, called a star cluster.

▲ This cluster of young stars, with many stars of different colours and sizes, will gradually drift apart, breaking up the cluster.

## QUIZ

1. What is a nebula?
2. How long has the Sun been shining?
3. What colour are large hot stars?
4. What is a group of new young stars called?

Answers:
1. a cloud of dust and gas in space 2. about 4 billion years 3. bluish-white 4. star cluster

# 48

**Smaller stars live much longer than huge stars.** Stars use up their gas to make energy, and the largest stars use up their gas much faster than smaller stars. The Sun is about half way through its life. It has been shining for about 5 billion years and will go on shining for another 5 billion years.

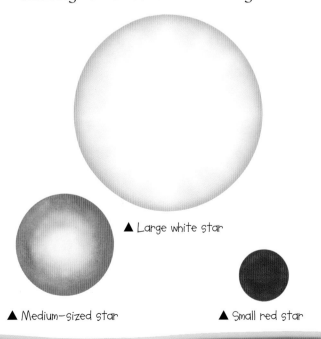

▲ Large white star

▲ Medium-sized star

▲ Small red star

# 47

**Large stars are very hot and white, smaller stars are cooler and redder.** A large star can make energy faster and get much hotter than a smaller star. This gives them a very bright, bluish-white colour. Smaller stars are cooler. This makes them look red and shine less brightly. Ordinary in-between stars like our Sun look yellow.

# Death of a star

**49** **Stars begin to die when they run out of gas to make energy.** The middle of the star begins to shrink but the outer parts expand, making the star much larger.

Supernova explosion

Ordinary star

▶ At the end of their lives stars swell up into red giant stars or even larger red supergiants.

Red giant star

**50** **Red giant stars are dying stars that have swollen to hundreds of times their normal size.** Their expanding outer layers get cooler, making them look red. When the Sun is a red giant it will be large enough to swallow up the nearest planets, Mercury and Venus, and perhaps Earth.

**51** **A red giant becomes a white dwarf.** The outer layers drift away, making a halo of gas around the star. The starlight makes this gas glow and we call it a planetary nebula. All that is left is a small, hot star called a white dwarf which cannot make energy and gradually cools and dies.

**52** Very heavy stars end their lives in a huge explosion called a supernova. This explosion blows away all the outer parts of the star. Gas rushes outwards in all directions, making a glowing shell. All that is left is a tiny hot star in the middle of the shell.

Black hole

◀ After a supernova explosion, a giant star may end up as a very tiny hot star or even a black hole.

Black dwarf star

White dwarf star

▲ When the Sun dies it will become 100 times bigger, then shrink down to 100 times smaller than it is now.

**I DON'T BELIEVE IT!**
Astronomers only know that black holes exist because they can see flickers of very hot gas near one just before they are sucked in.

**53** After a supernova explosion the largest stars may end up as black holes. The remains of the star fall in on itself. As it shrinks, its gravity gets stronger. Eventually the pull of its gravity can get so strong that nothing near it can escape. This is called a black hole.

# Billions of galaxies

**54** The Sun is part of a huge family of stars called the Milky Way Galaxy. There are billions of other stars in our Galaxy, as many as the grains of sand on a beach. We call it the Milky Way because it looks like a very faint band of light in the night sky, as though someone has spilt some milk across space.

▶ Seen from outside, our Galaxy would look like this. The Sun is towards the edge, in one of the spiral arms.

**55** Curling arms give some galaxies their spiral shape. The Milky Way has arms made of bright stars and glowing clouds of gas which curl round into a spiral shape. Some galaxies, called elliptical galaxies, have a round shape like a squashed ball. Other galaxies have no particular shape.

**I DON'T BELIEVE IT!**
If you could fit the Milky Way onto these two pages, the Sun would be so tiny, you could not see it.

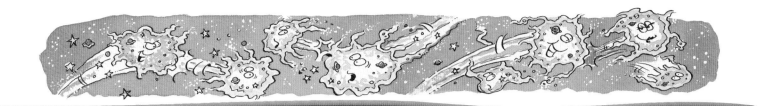

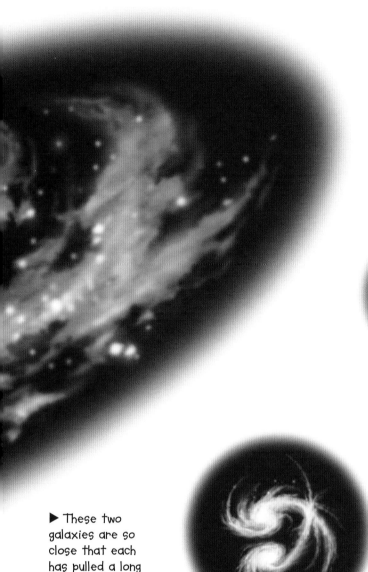

**56** There are billions of galaxies outside the Milky Way. Some are larger than the Milky Way and many are smaller, but they all have more stars than you can count. The galaxies tend to stay together in groups called clusters.

▲ A cluster of galaxies has many different types, with large elliptical and spiral galaxies and many small irregular ones.

▶ These two galaxies are so close that each has pulled a long tail of bright stars from the other.

▼ From left to right these are spiral, irregular, and elliptical galaxies, and a spiral galaxy with a bar across the middle.

**57** There is no bump when galaxies collide. A galaxy is mostly empty space between the stars. But when galaxies get very close they can pull each other out of shape. Sometimes they look as if they have grown a huge tail stretching out into space, or their shape may change into a ring of glowing stars.

# What is the Universe?

**58** **The Universe is the name we give to everything we know about.** This means everything on Earth, from tiny bits of dust to the highest mountain, and everything that lives here, including you. It also means everything in space, all the billions of stars in the billions of galaxies.

▼ 1. All the parts that make up the Universe were once packed tightly together. No one knows why the Universe started expanding with a Big Bang.

**59** **The Universe started with a massive explosion called the Big Bang.** Astronomers think that this happened about 15 billion years ago. A huge explosion sent everything racing outwards in all directions. To start with, everything was packed incredibly close together. Over time it has expanded (spread out) into the Universe we can see today, which is mostly empty space.

▼ 2. As everything moved apart in all directions, stars and galaxies started to form.

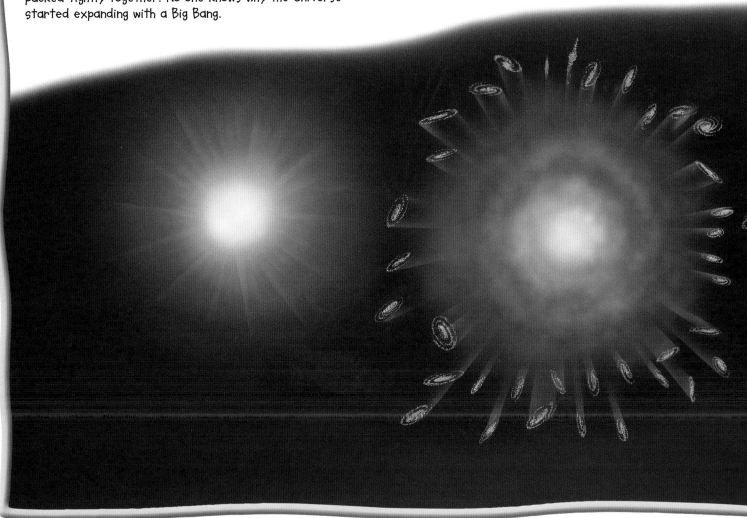

# 60

**The galaxies are still racing away from each other.** When astronomers look at distant galaxies they can see that other galaxies are moving away from our galaxy, and the more distant galaxies are moving away faster. In fact all the galaxies are moving apart from each other. We say that the Universe is expanding.

# 61

**We do not know what will happen to the Universe billions of years in the future.** It may keep on expanding. If this happens old stars will gradually die and no new ones will be born. Everywhere will become dark and cold.

▼ 3. Today there are galaxies of different shapes and sizes, all moving apart. One day they may start moving towards each other.

# 62

**The Universe may end with a Big Crunch.** This means that the galaxies would all start coming closer together. In the end the galaxies and stars would all be crushed together in a Big Crunch, the opposite of the Big Bang explosion.

## DOTTY UNIVERSE

**You will need:**

a balloon

Blow up a balloon a little, holding the neck to stop air escaping. Mark dots on the balloon with a pen, then blow it up some more. Watch how the dots move apart from each other. This is like the galaxies moving apart as the Universe expands.

▼ 4. The Universe could end as it began, all packed incredibly close together.

# Looking into space

**63** People have imagined they can see the outlines of people and animals in the star patterns in the sky. These patterns are called constellations. Hundreds of years ago astronomers named the constellations to help them find their way around the skies.

Scorpion

Great Dog

Southern Cross

▲ If you live south of the Equator, these are the constellations you can see at night.

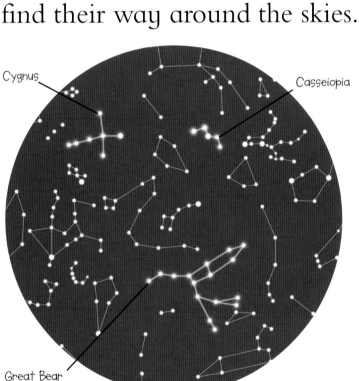

Cygnus

Casseiopia

Great Bear

▲ From the north of the Equator, you can see a different set of constellations in the night sky.

**64** Astronomers use huge telescopes to see much more than we can see with just our eyes. Telescopes make things look bigger and nearer. They also show faint, glowing clouds of gas, and distant stars and galaxies.

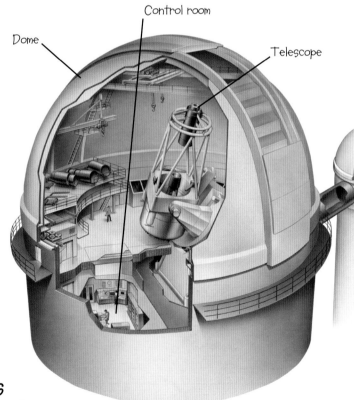

Control room

Dome

Telescope

▲ A huge dome protects this large telescope. It opens to let the telescope point at the sky, and both the dome and telescope can turn to look at any part of the sky.

▲ The Hubble Space Telescope takes much more detailed pictures and can see farther than any similar telescope.

**65** Space telescopes look even further to find exciting things in deep space. On Earth, clouds often hide the stars and the air is always moving, which blurs the pictures made by the telescopes. A telescope in space above the air can make clearer pictures. The Hubble Space Telescope has been circling the Earth for more than 10 years sending back beautiful pictures.

**66** Astronomers also look at radio signals from space. They use telescopes that look like huge satellite TV dishes. These make pictures using the radio signals that come from space. The pictures do not always look like those from ordinary telescopes, but they can spot exciting things that most ordinary telescopes cannot see, such as jets of gas from black holes.

## MOON-WATCH
**You will need:**
binoculars
On a clear night look at the Moon through binoculars, holding them very steady. You will be able to see the round shapes of craters. Binoculars are really two telescopes, one for each eye, and they make the Moon look bigger so you can see more detail.

▼ Radio telescopes often have rows of dishes like these to collect radio signals from space. Altogether, they act like one much larger dish to make more detailed pictures. The dishes can move to look in any direction.

# Three, two, one... Lift-off!

**67** To blast into space, a rocket has to travel nearly 40 times faster than a jumbo jet. If it goes any slower, gravity pulls it back to Earth. Rockets are powered by burning fuel, which makes hot gases. These gases rush out of the engines, shooting the rocket forwards.

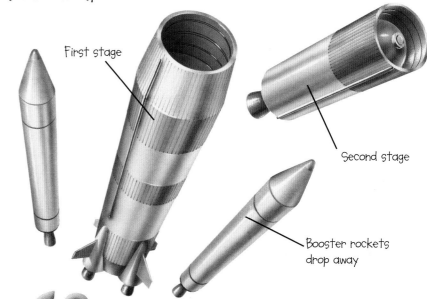

Satellite goes into space

Third stage

▶ Each stage fires its engine to make the rocket go faster and faster until it puts the satellite into space.

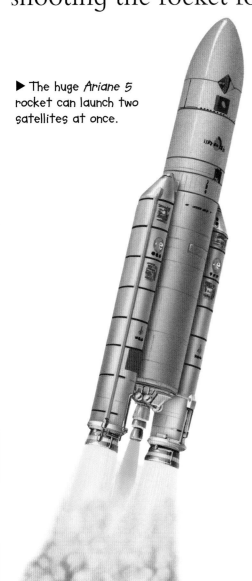

▶ The huge *Ariane 5* rocket can launch two satellites at once.

First stage

Second stage

Booster rockets drop away

**68** A single rocket is not powerful enough to launch a satellite or spacecraft into space. So rockets have two or three stages, which are really separate rockets mounted on top of each other, each with its own engines. When the first stage has used up its fuel it drops away, and the second stage starts. Finally the third stage takes over to go into space.

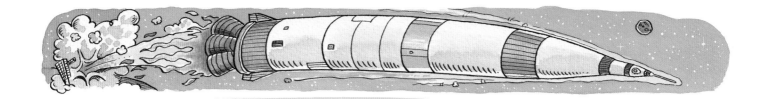

# 69

**The space shuttle takes off from Earth as a rocket.** It has rocket engines that burn fuel from a huge tank. But it also needs two large booster rockets to give it extra speed. The boosters drop away after two minutes, and the main rocket tank after six.

## ROCKET POWER

**You will need:**

a balloon

If you blow up a balloon and let it go, the balloon shoots off across the room. The air inside the balloon has rushed out, pushing the balloon away in the opposite direction. A rocket blasting into space works in a similar way.

# 70

**The shuttle lands back on Earth on a long runway, just like a giant glider.** It does not use any engines for the landing, unlike an aircraft. It touches down so fast, the pilot uses a parachute as well as brakes to stop it on the runway.

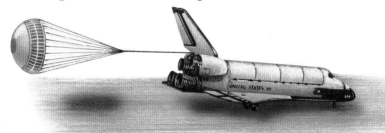

◀ The shuttle puts down its wheels and lands on the runway.

▲ The shuttle is blasted into space by three rocket engines and two huge booster rockets.

# Living in space

**71** **Space is a dangerous place for astronauts.** It can be boiling hot in the sunshine or freezing cold in the Earth's shadow. There is also dangerous radiation from the Sun. Dust, rocks and bits from other rockets speed through space at such speed, they could easily make a small hole in a spacecraft, letting the air leak out.

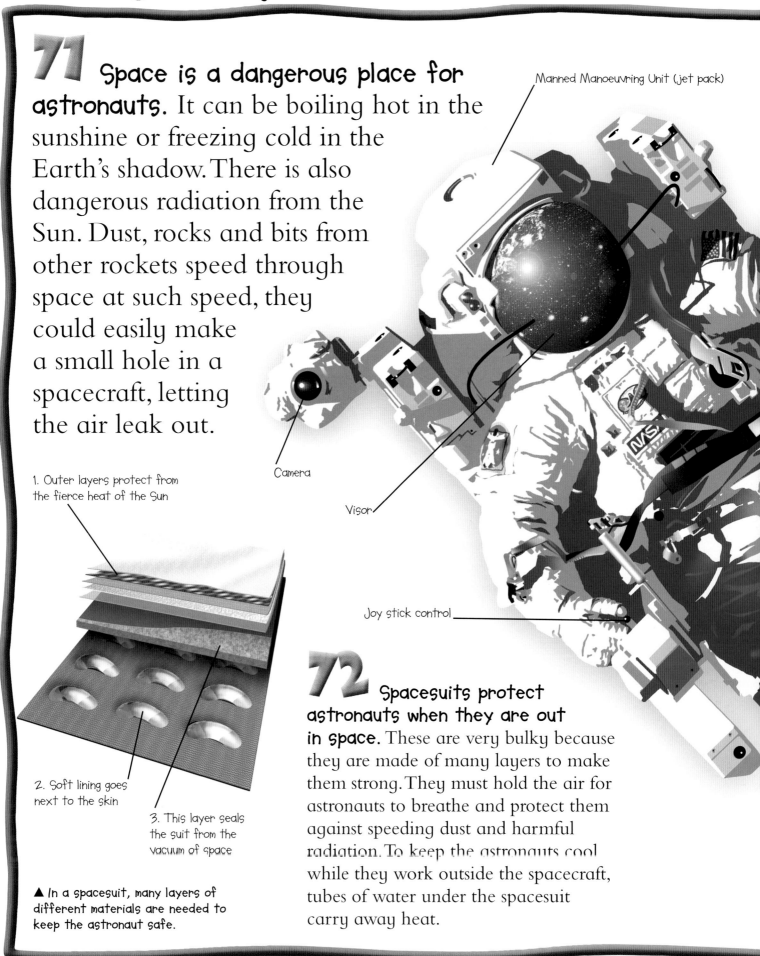

Manned Manoeuvring Unit (jet pack)

Camera

Visor

Joy stick control

1. Outer layers protect from the fierce heat of the Sun

2. Soft lining goes next to the skin

3. This layer seals the suit from the vacuum of space

▲ In a spacesuit, many layers of different materials are needed to keep the astronaut safe.

**72** **Spacesuits protect astronauts when they are out in space.** These are very bulky because they are made of many layers to make them strong. They must hold the air for astronauts to breathe and protect them against speeding dust and harmful radiation. To keep the astronauts cool while they work outside the spacecraft, tubes of water under the spacesuit carry away heat.

## SPACE MEALS

**You will need:**

dried noodles

Buy a dried snack such as noodles, that just needs boiling water added. This is the kind of food astronauts eat. Most of their meals are dried so they are not too heavy to launch into space.

Glove

Spacesuit

# 73 Everything floats around in space as if it had no weight. So all objects have to be fixed down or they will float away. Astronauts have footholds to keep them still while they are working. They strap themselves into sleeping bags so they don't bump into things when they are asleep.

esa

▲ Sleeping bags are fixed to a wall so astronauts look as though they are asleep standing up.

# 74 Astronauts must take everything they need into space with them. Out in space there is no air, water or food so all the things that astronauts need to live must be packed into their spacecraft and taken with them.

# Home from home

**75** A space station is a home in space for astronauts and cosmonauts (Russian astronauts). It has a kitchen for making meals, and cabins with sleeping bags. There are toilets, wash basins and sometimes showers. They have places to work and controls where astronauts can check that everything is working properly.

Solar panels for power

**76** The International Space Station, ISS, is being built in space. This is the latest and largest space station. Sixteen countries are helping to build it including the US, Russia, Japan, Canada, Brazil and 11 European countries. It is built up from separate sections called modules that have been made to fit together like a jigsaw.

Docking port

### I DON'T BELIEVE IT!
The US space station *Skylab*, launched in 1973, fell back to Earth in 1979. Most of it landed in the ocean but some pieces hit Australia.

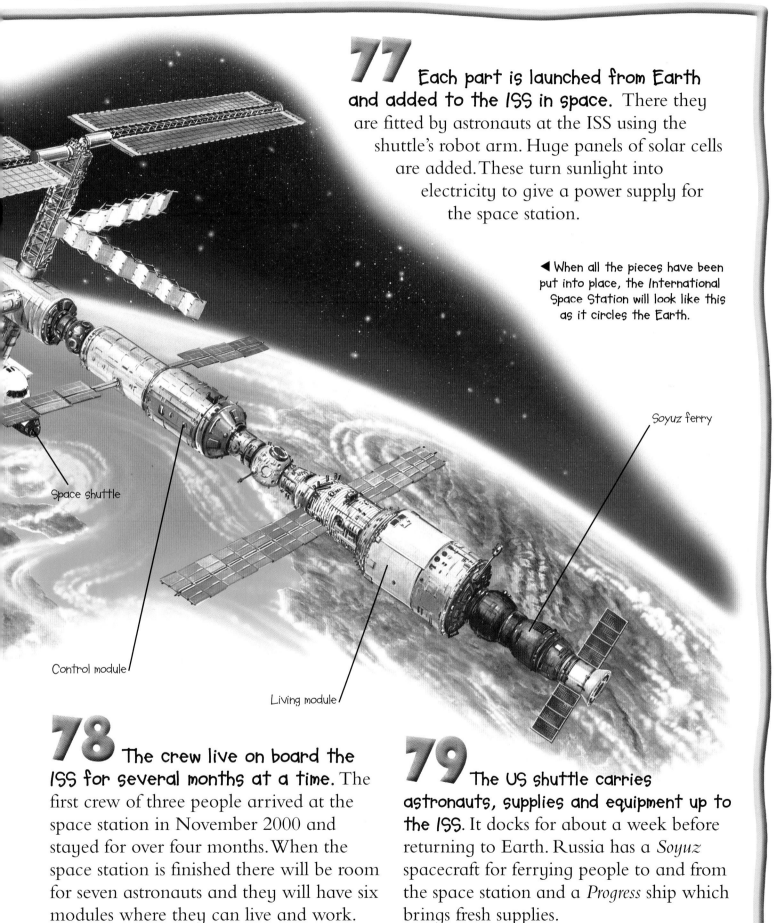

**77** Each part is launched from Earth and added to the ISS in space. There they are fitted by astronauts at the ISS using the shuttle's robot arm. Huge panels of solar cells are added. These turn sunlight into electricity to give a power supply for the space station.

◄ When all the pieces have been put into place, the International Space Station will look like this as it circles the Earth.

Soyuz ferry

Space shuttle

Control module

Living module

**78** The crew live on board the ISS for several months at a time. The first crew of three people arrived at the space station in November 2000 and stayed for over four months. When the space station is finished there will be room for seven astronauts and they will have six modules where they can live and work.

**79** The US shuttle carries astronauts, supplies and equipment up to the ISS. It docks for about a week before returning to Earth. Russia has a *Soyuz* spacecraft for ferrying people to and from the space station and a *Progress* ship which brings fresh supplies.

# Robot explorers

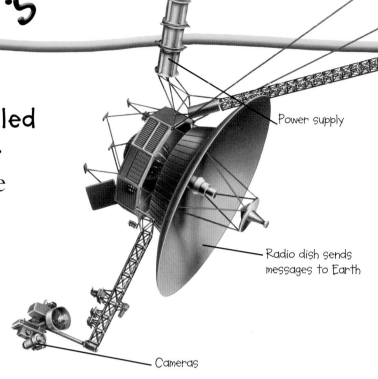

Power supply

Radio dish sends messages to Earth

Cameras

▲ Voyager 2 gave us close-up pictures of four different planets.

**80** Robot spacecraft called probes have explored all the planets. Probes travel in space to take close-up pictures and measurements. They send the information back to scientists on Earth. Some probes circle planets taking pictures. For a really close-up look, a probe can land on the surface.

**81** In 1976, two *Viking* spacecraft landed on Mars to look for life. They scooped up some dust and tested it to see if any tiny creatures lived on Mars. They did not find any signs of life and their pictures showed only a dry, red, dusty desert.

▼ The *Viking* landers took soil samples from Mars, but found no sign of life.

**82** Two *Voyager* probes left Earth in 1977 to visit the gas giant planets. They reached Jupiter in 1979, flying past and on to Saturn. *Voyager 2* went on to visit Uranus and then Neptune in 1989. It sent back thousands of pictures of each planet as it flew past.

▲ When *Galileo* has finished sending back pictures of Jupiter and its moons, it will plunge into Jupiter's swirling clouds.

# 83
**Galileo has circled Jupiter for more than six years.** It arrived in 1995 and dropped a small probe into Jupiter's clouds. Galileo sent back pictures of the planet and its largest moons. It was discovered that two of them may have water hidden under ice thicker than the Arctic ice on Earth.

◀ *Sojourner* spent three months on Mars. The small rover was about the size of a microwave oven.

# 84
**Mars Pathfinder carried a small rover called Sojourner to Mars in 1997.** It landed on the surface and opened up to let *Sojourner* out. This rover was like a remote control car, but with six wheels. It tested the soil and rocks to find out what they were made of as it slowly drove around the landing site.

# Watching the Earth

**85** **Hundreds of satellites circle the Earth in space.** They are launched into space by rockets and may stay there for ten years or more.

▶ Weather satellites look down at the clouds and give warning when a violent storm is approaching.

**86** **Communications satellites carry TV programmes and telephone messages around the world.** Large aerials on Earth beam radio signals up to a space satellite which then beams them down to another aerial, half way round the world. This lets us talk to people on the other side of the world, and watch events such as the Olympics Games while they are happening in faraway countries.

▼ Communications satellites can beam TV programmes directly to your home through your own aerial dish.

**87** **Weather satellites help the forecasters tell us what the weather will be like.** These satellites can see where the clouds are forming and which way they are going. They watch the winds and rain and measure how hot the air and the ground are.

▶ The different satellites each have their own job to do, looking at the Earth, or the weather, or out into space.

**88** **Earth—watching satellites look out for pollution.** Oil slicks in the sea and dirty air over cities show up clearly in pictures from these satellites. They can help farmers by watching how well crops are growing and by looking for pests and diseases. Spotting forest fires and icebergs that may be a danger to ships is also easier from space.

▼ Pictures of the Earth taken by satellites can help make very accurate maps.

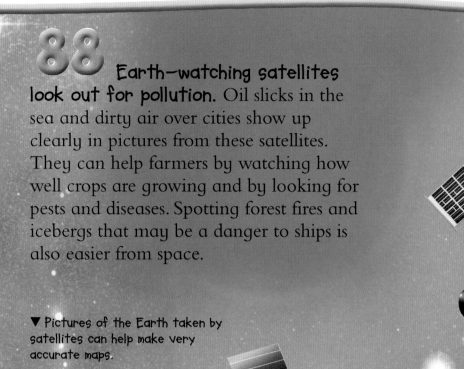

▲ Satellite telescopes let astronomers look far out into the Universe and discover what is out there.

**89** **Satellite telescopes let astronomers look at exciting things in space.** They can see other kinds of radiation, such as x-rays, as well as light. X-ray telescopes can tell astronomers where there may be a black hole.

### I DON'T BELIEVE IT!

Spy satellites circling the Earth take pictures of secret sites around the world. They can listen to secret radio messages from military ships or aircraft.

# Voyage to the Moon

**90** The first men landed on the Moon in 1969. They were three astronauts from the US *Apollo 11* mission. Neil Armstrong was the first person to set foot on the Moon. Only five other *Apollo* missions have landed on the Moon since then.

**91** A giant *Saturn 5* rocket launched the astronauts on their journey to the Moon. It was the largest rocket that had ever been built. Its three huge stages lifted the astronauts into space, and then the third stage gave the spacecraft an extra boost to send it to the Moon.

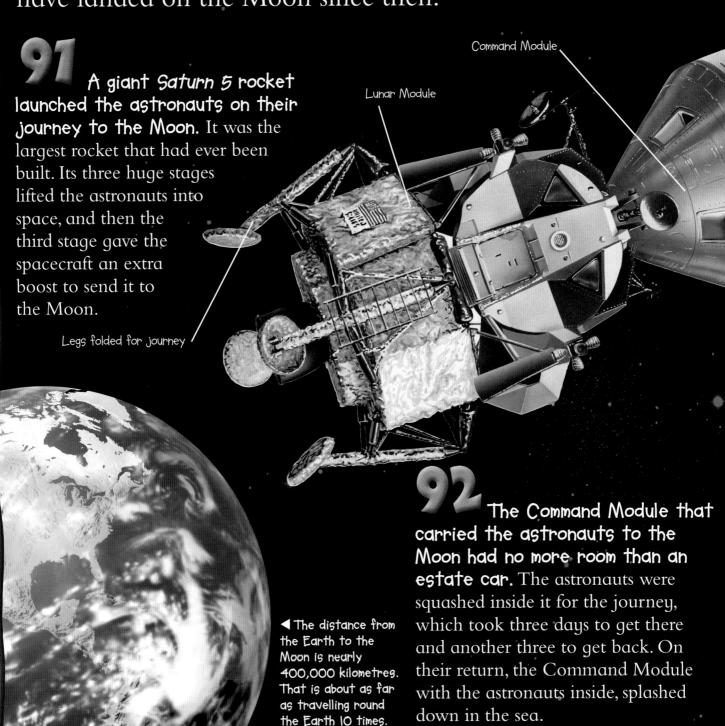

Command Module

Lunar Module

Legs folded for journey

◄ The distance from the Earth to the Moon is nearly 400,000 kilometres. That is about as far as travelling round the Earth 10 times.

**92** The Command Module that carried the astronauts to the Moon had no more room than an estate car. The astronauts were squashed inside it for the journey, which took three days to get there and another three to get back. On their return, the Command Module with the astronauts inside, splashed down in the sea.

▶ The longest time that any of the *Apollo* missions stayed on the Moon was just over three days.

Thrusters

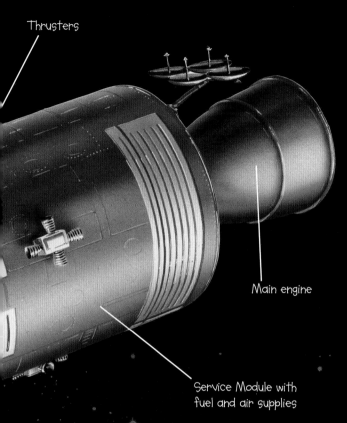

Main engine

Service Module with fuel and air supplies

▲ The Lunar and Command Modules travelled to the Moon fixed together, then separated for the Moon landing.

# 93

The Lunar Module took two of the astronauts to the Moon's surface. Once safely landed they put on spacesuits and went outside to collect rocks. Later they took off in the Lunar Module to join the third astronaut who had stayed in the Command Module, circling above the Moon on his own.

# 94

The Lunar Rover was a moon car for the astronauts to ride on. It looked like a buggy with four wheels and two seats. It could only travel about as fast as you can run.

# 95

No one has been back to the Moon since the last *Apollo* mission left in 1972. Astronauts had visited six different places on the Moon and brought back enough Moon rock to keep scientists busy for many years. Maybe one day people will return to the Moon and build bases where they can live and work.

## I DON'T BELIEVE IT!
On the way to the Moon an explosion damaged the *Apollo* 13 spacecraft, leaving the astronauts with little heat or light.

# Are we alone?

**96** The only life we have found so far in the Universe is here on Earth. Everywhere you look on Earth from the frozen Antarctic to the hottest, driest deserts, on land and in the sea, there are living things. Some are huge, such as whales and elephants and others are much too small to see. But they all need water to live.

▲ On Earth, animals can live in many different habitats, such as in the sea, the air, in deserts and jungles, and icy lands. How many different habitats can you see here?

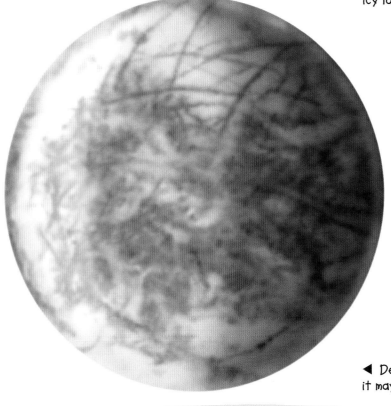

**97** There may be an underground ocean on Europa, one of Jupiter's moons. Europa is a little smaller than our Moon and is covered in ice. However, astronomers think that there may be an ocean of water under the ice. If so, there could be strange living creatures swimming around deep underground.

◄ Deep beneath the cracked, icy surface of Europa, it may be warm enough for the ice to melt into water.

## 98

Astronomers have found signs of planets circling other stars, but none like the Earth so far. The planets they have found are large ones like Jupiter, but they keep looking for a planet with a solid surface which is not too hot or too cold. They are looking for one where there might be water and living things.

▲ No—one knows what other planets would be like. They could have strange moons or colourful rings. Anything that lives there might look very strange to us.

## 99

Mars seems to have had rivers and seas billions of years ago. Astronomers can see dry river beds and ridges that look like ocean shores on Mars. This makes them think Mars may have been warm and wet long ago and something may have lived there. Now it is very cold and dry with no sign of life.

**I DON'T BELIEVE IT!**
It would take thousands of years to get to the nearest stars with our present spacecraft.

◄ This message could tell people living on distant planets about the Earth, and the people who live here.

## 100

Scientists have sent a radio message to a distant group of stars. They are hoping that anyone living there will understand the message about life on Earth. However, it will take 25,000 years to get to the stars and another 25,000 years for a reply to come back to Earth!

# Index